The Moon as seen from different latitudes

Revised 2nd edition

Peter D. Geldart
Member, RASC

The Moon as seen from different latitudes,
Revised 2nd edition.
Peter D. Geldart
member, RASC.
geldartp@gmail.com

c. 4100 words.
42 pages
4" x 6"

Arial 8
Courier New 14
Times New Roman 11

Cover:
A gibbous Moon rising over a lake on a December evening (note ice in the distance). Looking southeast from 45.4693 N latitude, 75.8106 W longitude. Author's photograph c. 1990.

ISBN 978-1-998321-38-4

Petra Books
MBO Coworking
78 George St., Suite 204
Ottawa ON K1N 5W1
613-294-2205

Previously published, in part, in the British Astronomical Association Journal, April 2025.

Contents

Abstract

The Moon's altitude above the horizon depends on your latitude and the angle the Moon's orbit makes with the Earth's equatorial plane (its declination). The formula for maximum altitude is given. A creature of the tropics, the Moon can only be seen at the zenith within, at the most, 28.5° latitude north and south. The author presents charts of the Moon's altitude as seen from various latitudes in summer and winter and discusses upper and lower transits.

Geldart

Introduction

This essay aims to highlight the factors influencing the Moon's apparent path and altitude when observed from different latitudes. It is the same Moon in the same phase presenting itself to everyone on the night side of Earth whatever their latitude. The Moon may also be seen in daytime situations such as having a pale Moon in the western sky as the Sun climbs in the east, or the full Moon rising in the east as the Sun sets in the west.

The charts on the following pages show the curves of altitude of the Moon as seen from three low-to-mid latitudes of 0° (the equator), 22° and 45° and three high latitudes of 70°, 80° and 90° (the pole). To give context, populated places at these latitudes include Rio de Janeiro and Singapore (0°), Hong Kong and Sao Paulo (22° N and S), Venice and Queenstown (45° N and S), Inuvik and Murmansk (70° N), and Alert (80° N); the only occupied place at a pole is the Amundsen–Scott South Pole Station (90° S).

Because of the Earth's eastward rotation, we see the Moon rise in the east, transit (looking towards the equator), and set in the west.[1] As with the Sun, planets and stars, the Moon's motion to the west is illusory: it is the observer who is being whisked along eastward by the Earth's rotation. The Moon's apparent westward progress is a bit less than the background stars because of it's own real eastward orbit.[2]

I have used data from NASA's JPL Horizons[3] with a longitude of Greenwich (0°), Universal Time (UT) and the sample year 2030.

---

1 Transit is when a celestial object appears to cross the observer's meridian, an imaginary line running from one pole to the other through the observer's zenith directly overhead. The terms "rise, transit, set" (RTS) are artificial terms for the effect of the Earth's rotation. See time lapse by Aryeh Nirenberg at https://youtu.be/1zJ9FnQXmJl
2 The Moon's orbit to the east "averages 3,681 kilometres per hour…corresponding to a mean angular velocity in the celestial sphere of about 33' [arcminutes] per hour…[by coincidence its] apparent diameter." *The Moon, Our Nearest Celestial Neighbour*. Zdeněk Kopal, p6, Chapman and Hall, London, 1960.
3 The NASA JPL Horizons data service at
https://ssd.jpl.nasa.gov/horizons/
Other sites of interest include
- the United States Naval Observatory data service at
https://aa.usno.navy.mil
  - Time and Date at https://www.timeanddate.com/moon/

Methodology

I started this investigation intrigued by the fact that the speed of eastward rotation of a point on the Earth's surface declines as latitude increases, and the celestial sphere appears to move more slowly to the west until as viewed from the pole the stars are circumpolar. The Moon, whose orbit is prograde, appears to move eastward against the background stars by 13.2° per day[4]. My hypothesis was that the Moon's apparent westward movement should lessen as latitude increases and near and at the pole it should be seen to move east in its true orbit.

Examining the Moon's ephemerides at JPL Horizons in detail (right ascension, azimuth, local apparent angle hour, sky motion[5]) I could not find a factor which declines as the latitude of the observer increases.

However, the Moon does stay above the horizon for multiple days at high latitudes, and this must be related to the shorter circumference

---

4 https://public.nrao.edu/ask/variability-of-the-moons-apparent-motion-through-the-sky/
5 JPL Horizons settings: R.A._(a-app), dRA*cosD, Azi_(a-app), dAZ*cosE, L_Ap_Hour_Ang, Sky_motion, Sky_mot_PA, and RelVel-ANG.

and lower rotational speed. I also discovered a few dates in the sample year (2030) at which, at 90°, the Moon rose at a western azimuth and set in the east. But there were many seemingly random rise and set azimuth numbers.

Jeff C., developer of the Sunmooncalc ephemeris, who also brought Equation (1) to my attention, as well as the references to Duffett-Smith and Meeus, advised:

> "...the relative contributions[6] don't change with latitude. At the poles, lineal speed is zero and direction essentially meaningless. ... At extreme latitudes, rise and set are determined primarily by changes in declination, so the azimuth appear to be somewhat random. ... The rate of change depends on declination as well as latitude, and there's no simple formula like the one for maximum altitude".
>
> - Jeff C., email communication, 2025

Concerning the apparent motion of the

---

6 "A sidereal day is 23h 56m 4s ... so Earth's angular speed is $\omega_E = 360°/23.934444$ hr = $15.041085°$/hr. The Moon completes an orbit in a sidereal month, so its orbital angular speed is $\omega_M = 360°/27.321661$ days = $0.54901494°$/hr. Because the Moon's orbit is prograde, the Moon's angular speed with respect an observer on Earth is $\omega_E - \omega_M = 15.041085°$/hr $- 0.54901494°$/hr = $14.49207°$/hr. So 96.3% of the motion is due to Earth's rotation." — Jeff C., email communication, 2025.

Moon as seen from different latitudes, Jon G. of JPL Horizons said:

> "Azimuth & elevation are local coordinates carried along by the rotation of the Earth, being based on local zenith direction and the plane perpendicular to it. … set the Moon (301) as a target, request output of quantity #2 (RA & Dec), #3 (RA & DEC rates), #4 (az-el angles), #5 (az-el rates), and/or #47 (sky motion)."
>
> - Jon G., email communication, 2025.

I was not able to show that the true orbit of the Moon begins to be revealed to observers as their latitude increases. Perhaps actual observations at these high latitudes to time the path of the Moon rather than relying on calculated data tables would provide an answer.

For the rest of the essay it was straight-forward to produce graphs in Microsoft Excel using the JPL Horizons data showing the apparent altitude of the Moon in winter and summer as seen from six sample latitudes.

A coordinate system

Like the ancients we can imagine a celestial dome above with pinpricks of light. On this is projected the Earth's lines of longitude and latitude.

Coordinate systems assist in understanding the Earth-Moon relationship. Duffett-Smith:

> "To fix the position of any astronomical object we must have a frame of reference, or coordinate system, which assigns a different pair of numbers to every point in the sky. The two numbers, or coordinates, usually refer to 'how far round' and 'how far up', just as do the longitude and latitude of an object on the Earth's surface. There are … the horizon system, the equatorial system, the ecliptic system and the galactic system." [7]

A longitudinal line from one pole to the other passing through the zenith directly above is the observer's meridian. As the Earth turns a celestial body appears to move east to west across the observer's meridian, when it will be at its highest altitude. This is its upper transit. Twelve hours later, as the Earth rotates and moves the observer to the "other side", it appears to cross the meridian again at its lower transit,

---

7 *Practical Astronomy with your Calculator.* Peter Duffett-Smith. Cambridge University Press, 2nd ed. 1981.

likely below your horizon, unless, at high latitudes, looking towards the pole, you see it as circumpolar, staying above your horizon.

A formula for the Moon's altitude can be derived. The Moon's maximum altitude, h_{max}, is calculated from its declination (δ) and the observer's latitude (ϕ) as follows:[8]

$$h_{max} = 90° - |\delta - \phi| \qquad \text{(Equation 1)}$$

Note that the altitude and declination numbers obtained from JPL Horizons are *topocentric* (the observer is on the Earth's surface):

> "For objects in the solar system … parallax is the difference in direction between a topocentric observation (by the actual observer at the Earth's surface) and a hypothetical geocentric observation [an observer at the centre of the Earth]." [9]

---

8 Also see Krisciunas K. et al. *The first three rungs of the cosmological distance ladder*, Am. J. Phys., 80(5), p. 430 (2012). https://scispace.com/pdf/the-first-three-rungs-of-the-cosmological-distance-ladder-1zeg8nff9i.pdf
9 Meeus J., *Astronomical Algorithms*, 2nd ed., Willmann-Bell Inc., Richmond, Virginia, 1988, p. 412.

1	2	3	4
Observer latitude on the Earth (deg)	Earth circumference (km)	Observer on the Earth's surface: linear speed of eastward rotation (km/hr) $2\pi R \times \cos(\text{lat}) / 24$ hr	Moon above the horizon when on the night side of Earth (hrs)
0° (equator)	40,000 km	1670 km/hr	12 hrs
22°	37,000	1550	6-12 hrs
45°	28,000	1200	6-12 hrs
70°	14,000	570	Various hrs and one 6-day period /month
80°	7,000	290	Various hrs and one 11-day period /month
90° (poles)	0	0	One 14-day period /month. (half a month)

Table 1 Variation in factors due to the Earth's eastward rotation.
Sources: https://www.vcalc.com/wiki/MichaelBartmess/Rotational-Speed-at-Latitude.
NASA JPL Horizons data service at https://ssd.jpl.nasa.gov/horizons/.

The Earth's rotation

The Sun, Moon, planets and celestial sphere overall appear to move from east to west because of the Earth's eastward rotation. It is common experience that the Moon and Sun appear to rise and set faster and more perpendicular to the horizon at the equator than at other latitudes. In addition, the speed to the east of an observer on the Earth's surface slows as latitude increases because the circumference to travel around in 24 hrs is less. At increasing latitudes the Sun and Moon rise/set at an angle to the horizon and take longer doing so. Above about 70° the Moon lingers above the horizon for multiple days because it is seen, in the N. hemisphere, to the south (upper transit) and continues to remain above the horizon as the observer rotates around the pole, seeing the Moon over the pole to the north in lower transit.

In Table 1 (left) the multi-day periods in column 4 at the three high latitudes must be related to the diminishing speed of rotation (column 3). Recall that in Summer at high latitudes the Sun is continuously above the horizon (midnight sun) so the view of the Moon may be attenuated.

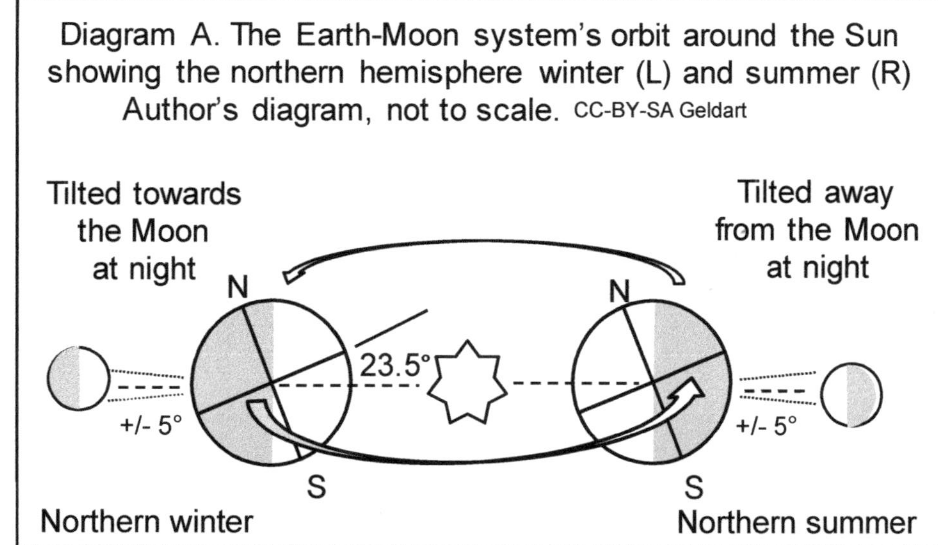

Diagram A. The Earth-Moon system's orbit around the Sun showing the northern hemisphere winter (L) and summer (R) Author's diagram, not to scale. CC-BY-SA Geldart

The Earth's tilt

As shown in Diagram A, the Earth is tilted on its axis by 23.5° such that in the northern winter (L) the northern hemisphere is tilted away from the Sun. Six months later the northern hemisphere is tilted toward the Sun giving the northern summer (R).[10]

Because the Sun and full Moon, as shown, are by definition opposite each other, when the Sun's declination is at a minimum in the northern winter (L), the full Moon's declination must be at a maximum, and vice versa in the northern summer (R). Accordingly, the full Moon's maximum altitude is greater in the winter than in the summer.

The varying inclination of the Moon's orbit off the ecliptic by about 5° is also shown.

---

10 Earth's axial tilt of 23.5° is the same throughout its orbit and only changes a few degrees over about 26,000 years as the orientation of its axis slowly rotates, or precesses, like that of a spinning top. See https://space-geodesy.nasa.gov/multimedia/videos/EarthOrientationAnimations/EOAnimations.html

The tropics

Since the Earth's axial tilt means the equator is inclined by about 23.5° to its orbit around the Sun, the *ecliptic,* the region in which the Sun can be at the zenith (its declination) ranges from 23.5° N to S. This is called the *tropics* (from the Greek *tropikós*, meaning 'turning'), and is bounded by the Tropic of Cancer (23.5° N) and the Tropic of Capricorn (23.5° S).

The Moon also has *lunar tropics* but they vary because of the Moon's 5° orbital tilt off the ecliptic, which, as a result of the precession[11] of the orbit, ranges from 18.5° up to a max. of 28.5° latitude N&S: above 28.5° N in the northern hemisphere the Moon is seen at midnight transit (as it crosses your meridian) to the south, and at latitudes greater than 28.5° in the southern hemisphere it is seen at transit to the north. The Moon can only be at the observer's zenith when its declination and the latitude of the observer are equal, meaning this only occurs up to a maximum of 28.5° latitude N-S.

---

11 The Moon's orbit precesses (rotates) over an 18.6-year cycle and the Moon's orbital tilt of 5° is either added to or subtracted from the Earth's tilt of 23.5° over this cycle, so that the Moon's inclination to Earth's equator varies between about 18.5° and 28.5° latitudes north-south.

The Moon's orbit is inclined to Earth's equatorial plane (by definition your horizon is parallel to the equator), so the Moon moves above and below this plane over the course of a lunar month. Because of this, the Moon's angle with the equator—its declination—varies over the course of the month. Jean Meeus:

> "The plane of the Moon's orbit forms with the plane of the ecliptic an angle of 5°. Therefore, in the sky the Moon is moving approximately along the ecliptic, and during each revolution (27 days) it reaches its greatest northern declination…and two weeks later its greatest southern declination. Because the lunar orbit forms with the ecliptic an angle of 5°, and the ecliptic an angle of 23° with the celestial equator, the extreme declinations of the Moon are between 18° and 28° (north or south), approximately." [12]

---

12 *Astronomical Algorithms*. 2nd ed. Jean Meeus. Willmann-Bell, 1998. *Note he has rounded off some figures.*

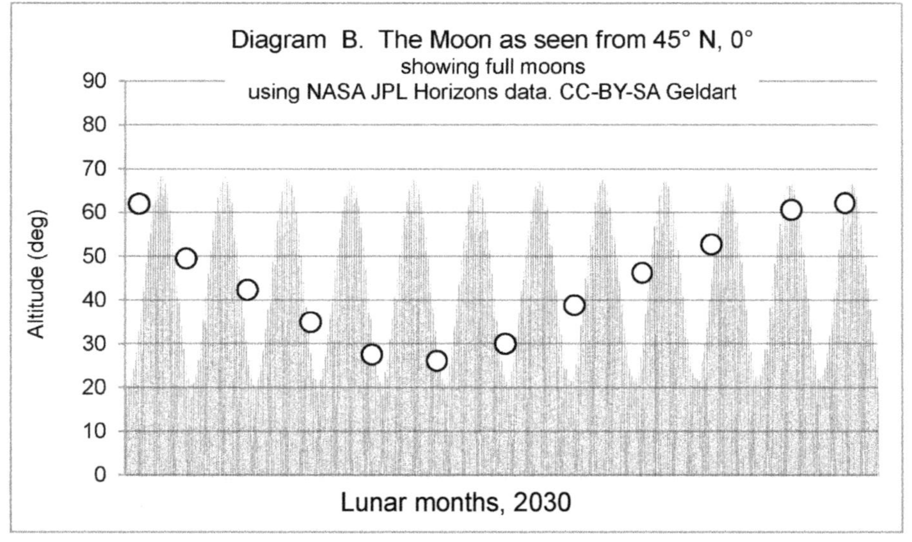

Diagram B. The Moon as seen from 45° N, 0°
showing full moons
using NASA JPL Horizons data. CC-BY-SA Geldart
Altitude (deg)
90
80
70
60
50
40
30
20
10
0
Lunar months, 2030

Lunar months

Plotting the Moon's altitude as seen from 45° N latitude, 0° longitude, for the whole of 2030 shows shaded undulations of sidereal lunar months of about 29.5 days which are similar all year with no seasonal variation (Diagram B). The Moon's orbit is independent of our seasons, our months, our diurnal day-night cycle, and its own phase[13], and for that matter the solstices and equinoxes of the Sun. The full Moons (when the Moon is opposite the Sun, more or less directly behind the Earth) are indicated and they are lower in summer and higher in winter due to the essentially fixed tilt of the Earth in its orbit (Diagram A).

The sample year 2030 is about midway in the precession of the Moon's orbit over 18.6

---

13 The Moon itself is always fully illuminated on its sunward side for its whole orbit (unless it happens to pass within the Earth's shadow) and it is only from Earth that we see the face that is towards us progressively illuminated in different phases. The convex curve of the illuminated portion faces the Sun, which is of course below the horizon at night. In daytime we may see a pale Moon (still nevertheless on the Earth's night side) with the Sun in the opposing part of the "dome of the sky". The Moon's phase is unrelated to its apparent path and altitude. It is only an artefact of illumination as seen by us on Earth.

years and its altitude varies by 5° over that period. The shaded curves would be about 5° less during the minor lunar standstill of 2015 and about 5° more during the major standstill of 2043. When the Moon is at its minimum (18.5°) and maximum (28.5°) declinations, it is called a *standstill* because the Moon rises at about the same point on the horizon for a few nights.[14] This can be called a lunistice (compare solstice, when the Sun is at the Tropic of Cancer at 23.5° N. or the Tropic of Capricorn at 23.5° S.).

---

14 See https://eprints.bournemouth.ac.uk/39590/

The Moon as seen from low-to-mid latitudes

The following Charts 1 and 2 show that on those dates the full Moon's altitude decreases as the observer's latitude increases ($0° \rightarrow 22° \rightarrow 45°$) and it appears higher in winter than in summer.

At latitudes below about 70° the Moon rises, transits and sets, and its subsequent lower transit 12 hours later is unseen below the horizon.

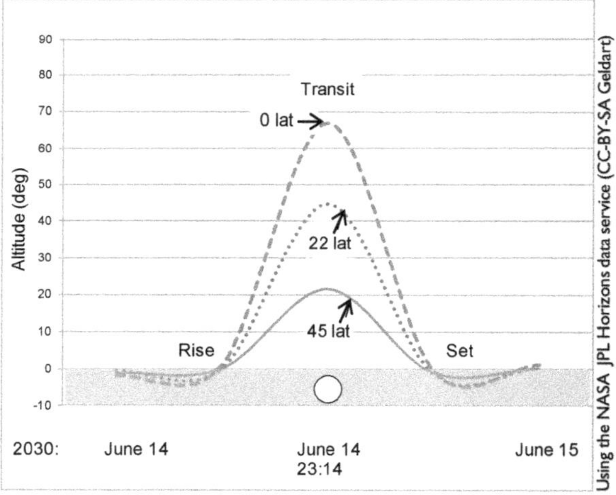

Note that the full Moon transits the observer's meridian (the direction towards the equator, that is, about due south for those in the N. hemisphere and due north from the S. hemisphere) about midnight on June 14 and a half-month later the new Moon (not illuminated on our side) transits about noon but the view is overwhelmed by sunlight (unless the Moon happens to pass in front of the Sun, giving a solar eclipse).

2. Full moon as seen from low latitudes in winter

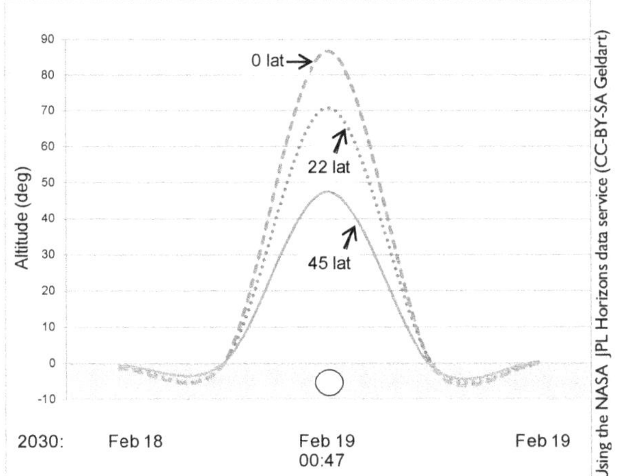

Chart 2 shows that the curves for the Moon's altitude are higher in February 2030 than in June (Chart 1).

3. Full moon as seen from low latitudes in winter

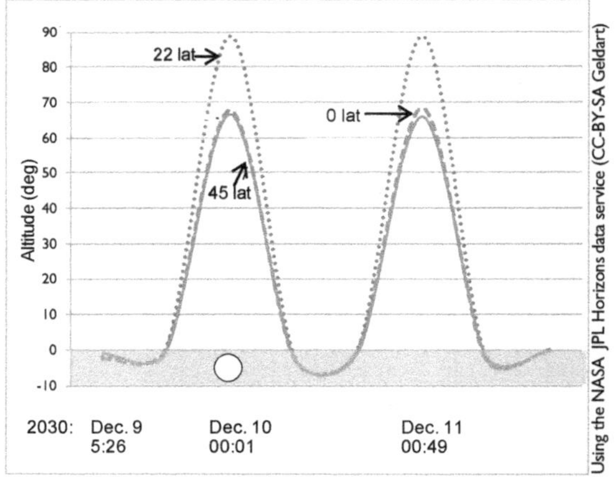

Just as well as being seen at the zenith from the equator, the Moon might instead be seen at the zenith from other latitudes, up to a maximum latitude of 28.5° N or S.

In Chart 3 for December 2030, the full Moon appears higher from latitude 22° than from the equator (0°) which was not the case in February when it was higher from 0° (Chart 2). The view from 0° and 45° are about the same but, in the N. hemisphere, from 0° the Moon is seen to the north and from 45° N. it is seen to the south.

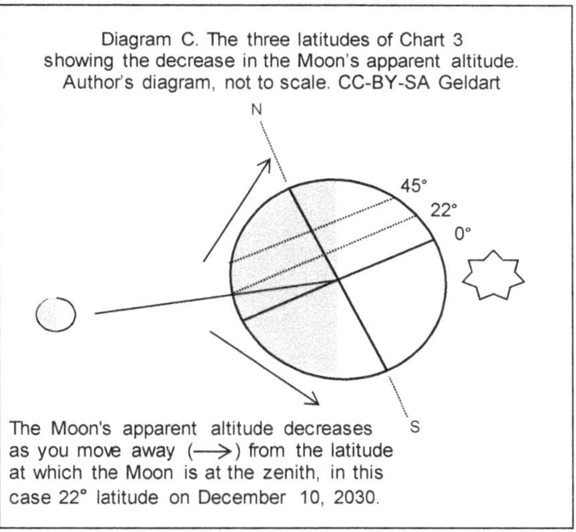

In support of Chart 3, Diagram C shows graphically that the Moon's apparent altitude is higher as seen from 22° latitude N. than from 0° (equator): it is at its maximum altitude close to the zenith.

This can be explained using Equation 1:

Full Moon as seen on December 10, 2030 (Chart 3)

$0°$ lat.: $h_{max} = 90° - |21° - \ \ 0°| = 69°$

$22°$ lat.: $h_{max} = 90° - |21° - 22°| = 89°$ (at the zenith)

$45°$ lat.: $h_{max} = 90° - |21° - 45°| = 66°$

Another way to consider this is by noting that, on this date, as seen from the equator the Moon appears to the north, from 22° N. latitude it appears directly overhead (approximately at the zenith), and from 45° N. latitude it appears to the south. When the observer's latitude (45°) is greater than the Moon's declination (c. 21°), the Moon's transit is to the south; when the observer's latitude (0°) is less, transit is to the north. Since the Moon is at the zenith as seen from 22° N latitude, all observers north of that see the Moon to the south, while those south of that see it to the north.

The Moon as seen from high latitudes

In the central area of the following Chart 4 (summer) in mid June it is clear that from 70° latitude the full Moon is barely seen at the horizon[15] and from 80° and 90° latitudes it has set.

---

15 Concerning the Moon near the horizon, refraction (the bending of light through the atmosphere causing celestial objects to appear higher) is taken into account by the NASA JPL Horizons data service. However, high land or clouds at the local horizon which may obscure a low Moon cannot be taken into account. Also, the observer's elevation above the ground is assumed to be zero, as if looking over extensive water or flat land.

4. Moon as seen from high latitudes in summer.

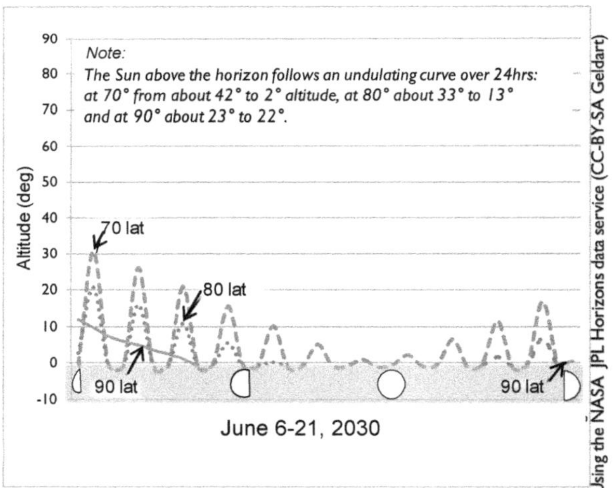

Above about 70° lat. in summer the Sun is starting to remain above the horizon for extended periods (midnight sun), with this duration increasing with the observer's latitude.

5. Moon as seen from high latitudes in winter.

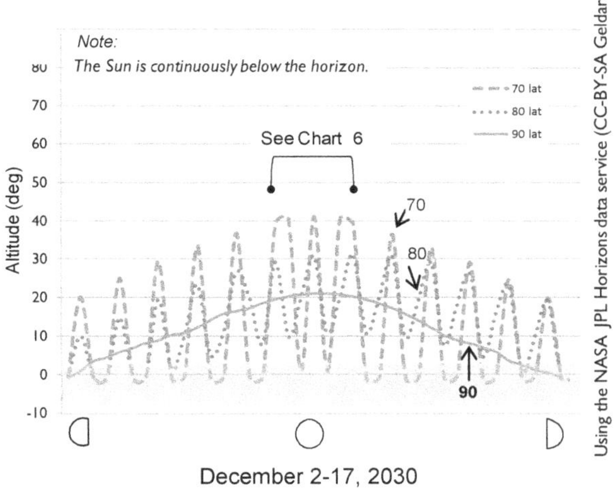

In Chart 5 in winter the undulating curves of the Moon are higher than in Chart 4 in summer due to the Earth's essentially fixed tilt (Diagram A). There are upper transits when the Moon reaches its highest altitude and crosses the observer's meridian, followed by lower transits 12 hours later when it has *not set* and crosses the meridian again. Note that the curve for 90° latitude is quite uniform because upper and lower transits are about the same. Thus the Moon is above the horizon and at low altitude for an extended period when it is, for about half a month, on the night side of the Earth. This is true for all latitudes above about 70° in winter: it remains above the horizon for about six days at 70°, eleven days at 80° and fourteen days (the whole half month) at 90°. The Moon has been undulating low in the sky the whole time.

Concerning the Sun, above about 66° latitude in winter it is below the horizon for increasingly longer periods as the observer's latitude increases (polar night).

6. Full moon as seen from high latitudes in winter (detail)

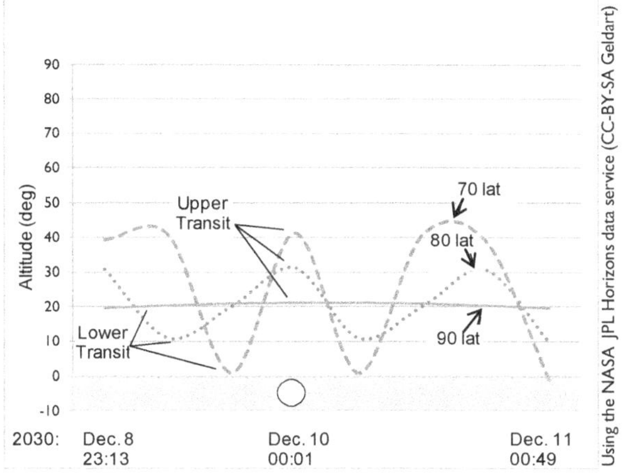

Zooming in on Chart 5, Chart 6 details the altitude of the full Moon for three days in December at high latitudes. Compare this to *low latitudes* in winter in which the curves are higher (Chart 2). At these high latitudes the upper and lower transits are all above the horizon. In the case of 90° the line is very flat because both transits are about the same altitude (20°, 21°).

At high latitudes, transits of the Moon crossing the observer's meridian are such that upper transits are seen looking to about 180° azimuth towards the equator, and 12 hours later when the observer is on the "other side" of the Earth's axis, lower transits are seen looking to about 0° azimuth over the pole. See Table 2 detailing transits for 70°, 80° and 90° N (northern hemisphere).

Notes to Table 2

In support of Chart 6.

Az ‡ For upper transits at these arctic latitudes observers are looking south to an azimuth of about 180°. Lower transits are seen to the north, looking back over the pole at an azimuth of about 0°. The reason the numbers in the Az ‡ column are not all exactly 0° and 180° is a matter of the precise minute-by-minute timing of the calculation in the JPL Horizon's ephemeris tables.

*** On these mid-winter dates the Moon is continuously above the horizon (there is no rise or set).

At 90° latitude (the pole) both Moon transits are about the same altitude (20°, 21°).

The values for altitude vary by 5° over the 18.6-year precession cycle of the Moon's orbit. For example, the 70° upper transit value of "41" would be about 5° less (in the mid 30s) at the minor lunar standstill of 2015, and about 5° more (in the mid 40s) at the major lunar standstill of 2043.

The Moon as seen from different latitudes

Table 2. Data for upper and lower transits of the Moon
as seen from high latitudes in winter.
CC-BY-SA Geldart, based on data from the
U.S. Naval Observatory and NASA's JPL Horizons

Year: 2030

Latitude: N 70 °

Date	Rise	Az.	Upper Transit.	Alt.	Az ‡	Set	Az.	Lower Transit.	Alt.	Az ‡
	h m	°	h m	°	°	h m	°	h m	°	°
Dec-08		***	23:07	41 South	182		***	10:43	1 North	1
Dec-09		***	23:55	41 South	181		***	11:31	1 North	1
Dec-10		***					***	12:20	1 North	0
Dec-11		***	00:44	41 South	182		***	13:08	0 North	0

Latitude: N 80 °

Date	Rise	Az.	Upper Transit.	Alt.	Az ‡	Set	Az.	Lower Transit.	Alt.	Az ‡
	h m	°	h m	°	°	h m	°	h m	°	°
Dec-08		***	23:07	31 South	182		***	10:43	10 North	0
Dec-09		***	23:55	31 South	181		***	11:31	11 North	1
Dec-10		***					***	12:20	11 North	0
Dec-11		***	00:44	31 South	182		***	13:08	10 North	1

Latitude: N 90 °

Date	Rise	Az.	Upper Transit.	Alt.	Az ‡	Set	Az.	Lower Transit.	Alt.	Az ‡
	h m	°	h m	°	°	h m	°	h m	°	°
Dec-08		***	23:07	21 South	181		***	10:43	20 North	2
Dec-09		***	23:55	21 South	180		***	11:31	21 North	1
Dec-10		***					***	12:20	21 North	2
Dec-11		***	00:44	21 South	180		***	13:08	20 North	1

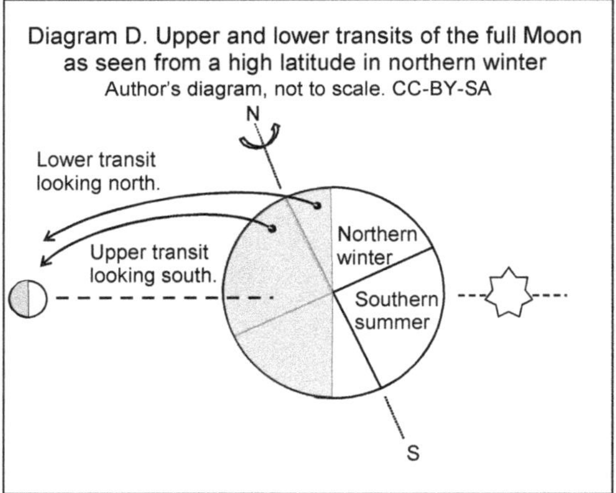

Diagram D depicts full Moon transits for someone in, for example, Alert, Canada at 80° latitude. Upper transit of the Moon is at about midnight as it crosses the observer's meridian above the horizon at about 180° azimuth (in the N. hemisphere looking south). As the Earth turns, about 12 hours later s/he reaches the "day" side (still in darkness) and see a lower transit to the north, looking back over the pole at about 0° azimuth.

Circumpolar

Throughout this time, and for the approx. 14 days the Moon is on the night side, it has, at latitudes above about 70°, been undulating above the horizon and is circumpolar: for 6 days as seen from 70° latitude, 11 days at 80° and the full 14 days, half a month, at 90°.

At high latitudes in summer both the Moon and Sun are circumpolar and they never set for extended periods. The Moon may at times be faint in the brighter sky.

At high latitudes in winter the Moon is circumpolar and the Sun is below the horizon.

Conclusion

The Moon's orbit is only dependent on its spacetime environment, i.e., its own mass and gravitational well, intermeshed with that of the Earth, Sun and the solar system as a whole. In graphs the apparent altitude of the Moon describes an undulating curve of constant shape following lunar months and spanning the years without regard to our daily rotation, our months, our seasons, the Sun's solstices and equinoxes, and the Moon's own phase. Yet its path above our horizon changes from night to night. This arises because the Moon orbits about 5° off the ecliptic and so its angle north or south of the Earth's equatorial plane (its declination) changes over the course of the lunar month. This declination, along with the latitude of the observer can be used to calculate the Moon's altitude as seen from any given location.

Two factors help in understanding the Moon's position. First, moving away from the tropical latitude at which the Moon is at the observer's zenith, it appears progressively lower in the sky. Second, due to Earth's (fixed) tilt, the full Moon appears higher during winter (when the Sun's declination is at its minimum and the Moon's is at its maximum) than in summer, when the situation is reversed, with the Sun at its maximum and the Moon at its minimum declination.

The observer should be able to understand the reasons for the Moon's position, and imagine what people are seeing at other latitudes.